FLUORSPAR MINING

Shawnee Books

FLUORSPAR MINING

PHOTOS FROM ILLINOIS AND KENTUCKY, 1905–1995

HERBERT K. RUSSELL

Southern Illinois University Press | Carbondale

Southern Illinois University Press
www.siupress.com

22 21 20 19   4 3 2 1

Cover illustration: *background*, purple fluorite (cropped; iStock, original image by MarcelC); *inset*, men at the Big Four Fluorspar Mine near Sheridan, Kentucky (courtesy Ben E. Clement Mineral Museum, Marion, Kentucky).

Frontispiece: Blue fluorspar is shown along with other minerals. Fluorspar is typically cube-shaped, occurs in many colors and countries, and is valued worldwide as a collectible and for its industrial and commercial uses. Photo by Steve Bonney.

Library of Congress Cataloging-in-Publication Data
Names: Russell, Herbert K., [date] author.
Title: Fluorspar mining : photos from Illinois and Kentucky, 1905–1995 / Herbert K. Russell.
Description: Carbondale : Southern Illinois University Press, [2019] | Series: Shawnee books | Includes bibliographical references.
Identifiers: LCCN 2018014174 | ISBN 9780809336685 (pbk.) | ISBN 9780809336692 (e-book)
Subjects: LCSH: Fluorspar—Illinois—Pictorial works. | Fluorspar—Kentucky—Pictorial works. | Mines and mineral resources—Illinois—Pictorial works. | Mines and mineral resources—Kentucky—Pictorial works. | Geology—Illinois—Pictorial works. | Geology—Kentucky—Pictorial works.
Classification: LCC TN948.F6 R78 2019 | DDC 622/.363—dc23
LC record available at https://lccn.loc.gov/2018014174

CONTENTS

ILLUSTRATIONS

ACKNOWLEDGMENTS

I am grateful to several individuals and institutions for their help in preparing this book. Julie Farley of the *Hardin County Independent* in Elizabethtown, Illinois, responded to my idea about a book of photos by forwarding a dozen pictures from her own files. Her "Out of the Past" columns complemented the photos with summaries of days gone by in the fluorspar-rich towns of Cave in Rock and Rosiclare, Illinois.

I am also indebted to the American Fluorite Museum in Rosiclare. Board president Eric Livingston, a geologist formerly with an Illinois fluorspar mine, read the manuscript, offered corrections, and granted permission to use the museum's photos. Linda Dutcher, a geologist previously with the same mine, also read the manuscript and made useful suggestions.

The Ben E. Clement Mineral Museum in Marion, Kentucky, offered a rich assortment of fluorspar mining photos in addition to its magnificent fluorspar collection. I am grateful to that museum's board, including director Ben E. Clement Jr. and William Frazer, for reading the manuscript and for permission to reproduce photos, and to Russ Adams for his assistance. Museum director Tina Walker provided help with photographs and useful information on several occasions. I appreciate her patience.

Thanks to the Illinois State Geological Survey: geologist F. Brett Denny read the manuscript with a critical eye and supplied especially valuable help with the introduction; Susan Krusemark and Anne Huber kindly reviewed photos in the survey's files and approved the use of several prints.

I am indebted also to the Special Collections Research Center in Morris Library at Southern Illinois University Carbondale for use of its photos of fluorspar miners in the Ben Gelman Collection. I appreciate the permission to reprint from the center's director, Pamela Hackbart-Dean, and the assistance of Aaron Lisec.

Steve Bonney provided color photos of fluorspar, including the frontispiece, and Bethany Belford took the photos of Rosiclare's double statue memorializing the fluorspar industry. Photographers Steve Buhman and Gary DeNeal made copies of photos in the Ben E. Clement Mineral Museum and the American Fluorite Museum, respectively. Other photo credits are shown with the captions.

Southern Illinois University Press's Barb Martin, Kristine Priddy, Linda Buhman, Amy Alsip, Wayne Larsen, and Lola Wilkening helped shape the book—the first pictorial history of fluorspar mining in the Illinois-Kentucky bistate area.

Acknowledgments

FLUORSPAR MINING

INTRODUCTION

The photographs presented in this book are from one of the world's richest deposits of fluorspar, a mineral so pretty that collectors seek it, and so useful that modern life benefits from it. It is known by different names to different people: fluorite to collectors and geologists, calcium fluoride (CaF_2) to chemists, and "spar" to those who mine it. Fluorspar helps prevent tooth decay and makes steel stronger, glass finer, and aluminum possible. In 1945, it helped transform uranium 238 into U-235 and end World War II with the atomic bomb.

There are hundreds of books devoted to fluorspar's beauty and many industrial uses, but there has not heretofore been a book of photographs devoted to those who mined it. Miners of all types, engineers, geologists, and state and regional historians will appreciate this book—and those who collect fluorspar may be surprised at the toil and danger that made this beautiful mineral available.

HISTORY OF THE FLUORSPAR INDUSTRY

Fluorspar's attractions have long been known. The Romans used it in metal work, and American Indians of the Mississippian culture shaped fluorspar into statuettes and jewelry in Tolu, Kentucky, as did those at Kincaid Mounds in Illinois. The modern fluorspar industry had its start in 1818, when explorer Henry Rowe Schoolcraft was traveling on the Ohio River and saw, near the Illinois "cliffs of the Cave-in-Rock," some "fine purple spars, and crystallized galena and other mineralized attractions."

Development in Kentucky dates to 1835–36, when President Andrew Jackson became co-owner of a mine northwest of Marion in Crittenden County;

in an era when men grew rich from land speculation, Old Hickory was involved in dozens of such transactions, including this one in search of the silver associated with lead (or *galena*, an important ore of lead) and fluorspar. Galena and fluorspar often occur together, and in the days before fluorspar's many uses were known, it was set aside while galena was processed into lead for bullets. After fluorspar's usefulness in purifying steel became better understood in the 1880s, these set-aside piles also started being processed. Kentucky's head start in mining resulted in it producing most of America's fluorspar by the 1890s.

The industry flourished in Illinois and Kentucky from the 1890s to the 1990s, and the area was, for a time, the most productive fluorspar mining center in the world. The surge in mining brought jobs and prosperity and shaped lives and landscapes accordingly. Unfortunately, the public is slowly forgetting this part of our past, so much so that a map is needed to show where these events took place.

For large sections of the twentieth century, Illinois-Kentucky deposits provided over 90 percent of the nation's fluorspar. Because the federal government classifies it as a strategic material, supplies were guarded during World War II and stockpiled during the Cold War.

In the mid-1950s, the U.S. government began lowering tariffs on foreign sources of fluorspar (a nod to the political clout of the American steel industry, which used fluorspar to help remove impurities from iron). Imports had always supplied some markets in the eastern United States, but greater volumes of foreign fluorspar from Mexico, Italy, and eventually China and Africa cut into the American market. Still, the fluorspar industry was big business throughout much of the twentieth century: Illinois made it the state mineral in 1965, and communities in both Illinois and Kentucky enjoyed its benefits. The Illinois-Kentucky deposits still supplied 75 percent of U.S. fluorspar needs in the 1970s, but cheaper foreign sources finally shut down Kentucky mines in 1985 and Illinois ones in 1995. The smaller, family-owned, mines also curtailed production.

Although substantial reserves of fluorspar remain in the Illinois-Kentucky Fluorspar District, the area seems likely to stay inactive until an international crisis revives it, or human ingenuity creates superior methods of extraction. But

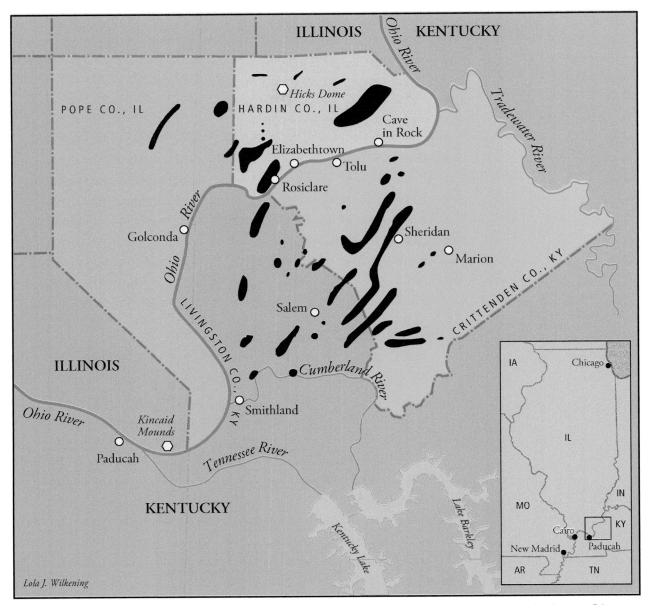

Shown in black, major deposits of the Illinois-Kentucky Fluorspar District are located on the lower Ohio River between the fabled outlaws' roost of Cave in Rock, Illinois, and Paducah, Kentucky. Some adjoining counties in Illinois and Kentucky also enjoyed significant deposits. Map adapted from information provided by Illinois State Geological Survey and Kentucky Geological Survey.

even if better times return, the industry and its workers will appear far different from this collection of photos. To maintain the photos' provenance, any relevant information printed, typed, or written on them has been retained and repeated in the captions as necessary for clarity.

USES AND APPLICATIONS OF FLUORSPAR

The many uses of fluorspar are determined by its purity or *grade*. Although there are variations in the classification of fluorspar, *acid grade* spar is more than 97 percent pure; *ceramic grade* is between 85 and 96 percent pure; and *metallurgical grade* is between 60 and 85 percent pure. In lay terms, heavyweight jobs involving metals such as iron have a greater tolerance for impurities than delicate ones involving the preparation of glass; enamel (sinks and bathtubs); and optics (microscopes, cameras, and telescopes). The highest grade is used to create hydrofluoric acid and products as diverse as dental porcelain and rocket fuel.

GEOLOGY OF THE ILLINOIS-KENTUCKY FLUORSPAR DISTRICT

Fluorspar's presence near the lower Ohio River is related to the region's frequent earthquakes. The New Madrid Fault and other fault systems penetrated the area for millions of years and created vertical, or mostly vertical, passageways into which gases and ore fluids were forced upward by volcanic forces. The fluorspar deposits' age and mode of emplacement are uncertain, but mineralized fluids carrying fluorine and other metals ascended through faults and underground fractures to form veins composed of fluorspar, with lesser amounts of minerals such as galena, sphalerite, barite, calcite, and quartz.

With few exceptions, most of the fluorspar in Kentucky is found in these vertical deposits in faults. Faults thus provide natural underground receptacles for ore fluids (liquefied minerals), which, when hardened, constitute veins and are said to be mineralized.

If the ore fluid's upward journey was blocked, it might move laterally into a layer of limestone. There the fluorine within the fluid combined with the calcium in the limestone to become *replacement fluorspar*, so named because it had replaced the limestone. Many of the immense deposits of fluorspar near Cave in Rock, Illinois, are of this easily mined horizontal or *bedded* type. It is here, too, that the most beautiful collectible specimens have been found.

One of the geologic curiosities in Illinois is Hicks Dome, an ancient *structural uplift* (or upheaval). An upward bulge in the earth in Hardin County, the dome is sometimes referred to as a *cryptovolcano*. Hicks Dome did not explode violently into the open air—it did not blow its top—but its location next to the fluorspar field is a sign of a geologically unstable area. Near Tolu, Kentucky, a smaller upheaval is known as the Tolu Arch. In other words, the geophysical phenomena that helped bring the riches of fluorspar close to the earth's surface also brought near volcanoes and ongoing concerns about earthquakes.

MINING IN THE ILLINOIS-KENTUCKY FLUORSPAR DISTRICT

The fluorspar outcrops that explorer Henry Rowe Schoolcraft saw glinting in the sun in 1818 were exploited, and miners later searched for the mineral by using trenches, test holes, and machine-driven core drills. The top producing counties were Hardin and Pope in Illinois, and Crittenden and Livingston in Kentucky. In Illinois, a farmer sinking a well near Rosiclare found galena and fluorspar deposits, and by 1843, miners were working at the nearby Blue Diggings Vein, named for the ore's bluish tint. Ore was shipped by wagon, barrel, barge, truck, and eventually by rail after the Illinois Central added a much-needed spur. The Illinois-Kentucky Fluorspar District was so rich that, at its peak, it was providing much of America's domestic needs and exporting large quantities abroad.

America's involvement in World War I in 1917–18 underscored the importance of fluorspar in steel production and national defense—and resulted in the creation of dozens of mines in addition to those in the four counties shown on the map on page 3. An eight-county area in western Kentucky had 103 mines and southeastern Illinois counted 86. Many major U.S. corporations were also represented, including Alcoa, Allied Chemical, Inland Steel, and U.S. Steel.

At Spar Mountain near Cave in Rock, a shovel operator removes overburden from a surface mine as a supervisor (*see arrow*) observes. The eponymously named Spar Mountain and nearby Lead Hill enjoyed great mineral wealth. The costs of operating a surface mine were about one-tenth of those of underground mining. Courtesy *Hardin County Independent*, Elizabethtown, Illinois.

THOMAS PARTAIN, MINE FOREMAN, BENZON FLOURSPAR CO'S. SPAR MOUNTAIN MINE NEAR CAVE-IN-ROCK, II ABOUT 1935

rods inserted deep into the tops of passageways and held in place with resin or glue at the top and metal flanges below. The latter made for a more spacious work area in which to maneuver large mining equipment. Both types of supports are shown on pages 58 and 68, respectively.

Wages for the post–World War II years—some of the industry's best—are summarized in the *Hardin County Independent* for July 22, 1948: underground workers at the Crystal Mine above would be paid $1.00 per hour; those who worked on the surface would receive 90 cents an hour—terrible wages by twenty-first century standards but a good living at a time when the national minimum wage was 40 cents an hour. Fluorspar is not classified as hard rock mining, but it is difficult to extract, and the usual means of dislodging it underground is with dynamite, a process that begins by drilling a cluster of holes several feet deep for placement of explosives. The explosives are inserted into the holes, tamped firmly into place with a wooden pole, and detonated.

The explosion yields a *muck pile*, which includes everything freed by the explosion—waste rock, clay, shale, minerals in addition to fluorspar, a great deal of dust, and the headache-inducing gas that accompanies exploding dynamite. The next step is to sort the muck pile's desirable materials from undesirable ones.

MILLING, CLEANING, AND CONCENTRATING

In the early decades of the industry, unwanted materials were discarded manually by *handpicking* and by moving the product about in water with rakes, shovels, and machine-driven washers. The best way to begin was by *milling*—crushing large pieces of ore into smaller ones, which could then be screened and sized. The crushed ore might be sent through a *spiral classifier*—an inclined turning spiral placed in water that carried heavy materials upward while fines settled to the bottom (simple but effective, one is shown on page 72). Alternately, materials might also be passed through a chemical slurry that caused heavy materials to settle to the bottom and lightweight ones to rise to the top—a process known as *heavy medium separation*, which is shown on page 76.

Cleaned fluorspar then traveled by barge up the Ohio River to steel-making towns such as Pittsburgh. Other barges made their way up the Mississippi River to steel facilities at Granite City, Illinois, or on to the Great Lakes and Chicago-area steel mills such as those in Gary, Indiana.

This process of cleaning and concentrating minerals (or *beneficiation*) was greatly improved in the 1920s with the advent of *froth flotation*. During this process, finely ground ore is mixed with water and chemicals to create a slurry. Different chemicals cause different minerals to rise to the surface, or *float*—hence the term froth flotation. The desired mineral remains in the froth on the surface where it can be isolated, as shown in a flotation mill at Minerva Mine #1 in Illinois on pages 10 and 78.

DANGERS OF FLUORSPAR MINING

In spite of improved cleaning processes, fluorspar mining remained a dangerous occupation. Underground workers, in particular, faced many hazards, including

Photos from Illinois and Kentucky, 1905–1995

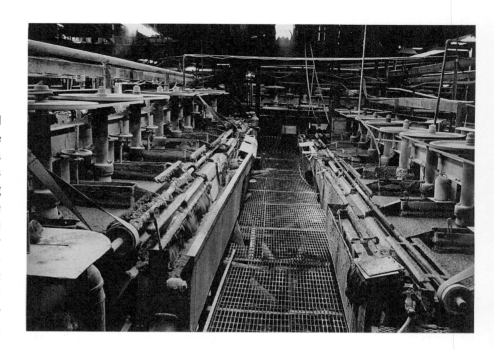

"View of Inside the Flotation Mill at Minerva Oil Company's Mine #1." Froth flotation machines fill a room at the Minerva Mine. Paddles sweep the cleaned particles riding on bubbles into a trough at the front of the machines. The process yields 97 percent pure fluorspar that is useful to the chemical industry as a component of hydrofluoric acid. A related photo is shown on page 78. Courtesy *Hardin County Independent*, Elizabethtown, Illinois.

crushing by roof falls; amputation of arms and fingers; accidents in the shaft; possible encounters with gas; and lung disorders. One of fluorspar's companion materials is silica, the dust of which can produce silicosis, a serious lung disease. This threat was ameliorated by adding hoses and sprays of water to "wet drills," but the milling process of crushing large pieces of ore into smaller pieces usually remained dusty, as did many of the industry's other jobs. Those afflicted with lung disorders were said to have "spar lung."

Flooding is also a danger in much deep mining. In Hardin County, much mining took place near the Ohio River, and at Rosiclare, some mining was done under the Ohio itself. Over the years, flooding caused several Illinois mines near the port town of Rosiclare to close, and even the well-known industrial service company of Halliburton was finally defeated by the Ohio.

In Kentucky, the worst flooding scare came in 1925, when five miners working at night at the hundred-foot level of the Hudson Mine near Salem were trapped as a water-filled pit above them suddenly caved in and cut off their exit. They scrambled to high spots in the mine and waited for six days, much of the

Fluorspar Mining

time in the dark with minimal food and water. Rescue equipment was delayed by rain and poor roads, and they had been given up for dead when found alive.

Water can also be a delivery system for deadly gas, as happened at the Ozark-Mahoning Barnett Mine in Illinois in 1971. Although local and state mine officials test for gas, seven miners perished at the eight-hundred-foot level some two thousand feet from the shaft when they were overcome by hydrogen sulfide gas, an uncommon phenomenon in fluorspar mines. Heavier than air, it

Miners Roger Marvel (*left*) and Ellis Jones (*right*) braved the dangers of darkness, drowning, entrapment, and crushing to help save five fellow miners in the 1925 Hudson Mine disaster in Salem, Kentucky. While rescuing the other miners, Marvel and Jones used lighting from their open-flame lamps. Courtesy Ben E. Clement Mineral Museum, Marion, Kentucky.

had been trapped in a pocket of water the miners had unknowingly broken into at the end of the previous work week. As the water soaked in or evaporated over the weekend, the pocket of gas remained in a low area. It overcame the men's respiratory systems when they returned at the beginning of the following week; two of the men lived long enough to tell what had happened.

MUSEUMS AND MEMORIALS

Today, only memories, museums, and commemorations remain of the bistate fluorspar industry. The end of World War II in 1945 brought with it a postwar prosperity in which both small and large pieces of fluorspar became popular with collectors, who usually refer to it as "fluorite." Three Illinois mines— Denton, Annabel Lee, and Minerva #1 (later Ozark-Mahoning)—produced some of the most colorful and sought-after specimens in a market that became worldwide.

In Marion, Kentucky, the successful mine owner-operator Ben E. Clement (1891–1980) established the mineral museum that bears his name. It holds one of the largest and finest collections of fluorspar specimens in the world and also houses samples of other minerals, fossils, and petrified wood. Tools, maps, letters, equipment, and records complement this extensive collection, along with informative photos, several of which appear within. Fluorspar crystals of specimen, or collectible, quality are on display and for sale along with cut and polished gemstones.

In Illinois, the American Fluorite Museum in Rosiclare is in the former corporate office of the Rosiclare Lead and Fluorspar Mining Company. In business from 1893 to 1954, it was once the largest fluorspar mine in the nation and, some said, the world. Many of the photos in this book are courtesy of this museum, which also houses mining artifacts such as protective and disaster gear, scale models of *headframes* (the structures that tower above mine shafts), and other accoutrements. Outside, ore cars bespeak the activities once conducted nearby, along with heavy steel buckets that alternately lowered men and materials and brought up ore. An annual Hardin County Fluorspar Festival is held in Rosiclare with pageants, parades, and memorials.

Above: Purple fluorspar is attached to a matrix of white calcite. The Ben E. Clement Mineral Museum in Marion, Kentucky, has one of the most expansive fluorspar collections in the world, with many specimens on display or for sale. Photo by Steve Bonney. *Left*: A curious turtle looks out from his place among locally made art objects at the American Fluorite Museum in Rosiclare, Illinois. Folk art and polished stones made of fluorspar have outlasted the industry from which they sprang. Photo by Gary DeNeal.

A serious-faced wife hands her husband his auxiliary lamp and a dinner pail in this detail from David Seagraves's 2017 statue in Rosiclare. Photo by Bethany Belford.

As this book was being prepared in 2017, at least one mine in Illinois was marketing ornamental fluorspar for the collector and gemstone markets, and at least one production mine was working in Kentucky on a vein with a reported width of sixty feet. But it was clear, too, that the old days shown in this book are gone. Underground mining was still a man's world when most of these photos were taken; only one female employee, a geologist working on the surface, is pictured (see the photo on page 84). But women were an integral part of the mining team, a fact acknowledged in Rosiclare's 2017 double statue of a miner and his wife shown in detail here and in full on page 83.

SOURCES

American Fluorite Museum, brochure, n.p., n.p., 1996, details of former headquarters of the Rosiclare Lead and Fluorspar Mining Company and the office building remaining as a museum.

Bailie, Harold, E. Powell, William Melcher, and F. J. Myslinski. *Fluorspar Mining Methods and Costs, Ozark-Mahoning Co., Hardin County, Ill.* ([Washington, D.C.]: U.S. Department of the Interior, Bureau of Mines, 1960), p. 19, special permit to use diesel motors underground; p. 26, diesel-powered generator used to hoist when electricity fails.

Bain, H. Foster. *The Fluorspar Deposits of Southern Illinois* (Washington, D.C.: Government Printing Office, 1905), summary of early mining and mines.

Bastin, Edson S. *The Fluorspar Deposits of Hardin and Pope Counties, Illinois* (Urbana, IL: State Geological Survey, 1931), p. 69, mining under the Ohio River.

Bates, Robert L. *Geology of the Industrial Rocks and Minerals* (New York: Dover, 1969), p. 278, Illinois-Kentucky District as most productive fluorspar region in the world.

Ben E. Clement Mineral Museum, brochure, n.p., n.p., n.d., photos of museum, founder Clement, fluorspar specimens, Crittenden County Historical Museum, and local heritage.

Bradbury, J. C., G. C. Finger, and R. L. Major. *Fluorspar in Illinois* (Urbana: Illinois State Geological Survey, 1968), geology, mining, milling, economic aspects, uses.

"Cave in Rock Hardin County Fluorite," web, most attractive fluorspar from the Minerva, Denton, and Annabel Lee Mines in Illinois, http://www.spiritrockshop.com/Fluorite_Hardin_County.html.

Clement, Ed. Conversations and manuscript corrections concerning Kentucky fluorspar mining, 2017.

Crittenden County, Kentucky, History and Families. Vol. 1, *1842–1891* (Marion, KY: Riverbend Publishing, 1991), pp. 423–30, details of mines and dangers; specifics of high-grade fluorspar from Holly Mine; and use of waste rock with fluorspar fragments on roads.

Denny, F. Brett, Illinois State Geological Survey, Champaign, IL. Manuscript suggestions and corrections, 2017.

Dutcher, Linda. Conversations and manuscript corrections concerning Illinois fluorspar mining, 2016, 2017.

Esters, Stephanie. "Town Unveils Mining Memorial," *Southern Illinoisan*, June 18, 2017, p. A3, David Seagraves's statue of miner and wife in Rosiclare, the "Fluorspar Capital of the World."

Farley, Julie. "Out of the Past," *Hardin County Independent*, April 21, 2016, p. 7, events of April 12, 1971, when hydrogen sulfide gas killed seven miners in the Barnett Mine Shaft of Ozark-Mahoning Mining.

———. "Out of the Past," *Hardin County Independent*, July 24, 2014, p. 7, events of July 22, 1948, when underground workers received a wage of one dollar per hour and surface workers ninety cents per hour.

———. "Out of the Past," *Hardin County Independent*, October 6, 2016, p. 7, sixty housing units under the name Spardale.

Frazer, William. Conversation and manuscript corrections concerning Kentucky fluorspar mining, 2016, 2017.

Hall, Edward Emerson. *The Geography of the Interior Low Plateau and Associated Lowlands of Southern Illinois* (St. Louis: J. S. Swift, 1940), p. 78, capacity of barges at 550 tons and Rosiclare Lead and Fluorspar Mining Company as largest such company in the world.

Hardin County Historical and Genealogical Society, *1999 Calendar*, back cover, author reprint from *Hardin County Independent* showing eighty-six fluorspar mines.

Kentucky Geological Survey, "Western Kentucky Fluorspar District" (Lexington: University of Kentucky, [ca. 2012]), web, p. 1, 75 percent of U.S. fluorspar from Illinois-Kentucky mines in 1970s and remaining substantial reserves, http://www.uky.edu/KGS/minerals/im_fluorspardistrict.php.

Kilgore, Catherine C., Sandra R. Kraemer, and James A. Bekkala. *Fluorspar Availability—Market Economy Countries and China: A Minerals Availability Appraisal*

([Washington, D.C.]: U.S. Department of the Interior, Bureau of Mines, 1985), decline of U.S. share of market.

Livingston, Eric. Conversations and manuscript corrections concerning Illinois fluorspar mining, 2016, 2017.

Mines and Minerals in Illinois: An Industrial Romance, compiled from records and surveys of State Department of Mines and Minerals [Springfield, IL: ca. 1945], pp. 90–93, fluorspar.

Schoolcraft, Henry Rowe. *The Indian in His Wigwam, or Characteristics of the Red Race of America* (New York: W. H. Graham, 1848), p. 23, "fine purple spars."

Schuberth, Christopher J. *A View of the Past: An Introduction to Illinois Geology* (Springfield: Illinois State Museum, 1986), pp. 106–19, overview of Illinois fluorspar.

Selected Areas within the Western Kentucky Fluorspar District and the Illinois Fluorspar District and a Tour of the Ben E. Clement Mineral Museum, compiled by Richard A. Smath, Kentucky Society of Professional Geologists fall field trip, October 23, 2010, web, p. 2, uranium 238 to U-235; p. 6, forty-five-foot-wide vein; p. 14, sixty-foot-wide vein; appendix, 103 mines in Kentucky.

Smith, Wes. "A Real Rock of the Crystal Trade," *Chicago Tribune*, December 17, 1990, web, fluorspar in lunch buckets and coolers, http://articles.chicagotribune.com/1990-12-17/features/9004140402_1_spar-mineral-illinois/2.

Through the Years: The 47th Annual Fluorspar Festival . . . , pamphlet, n.p., n.p., 2011, overview of Hardin County fluorspar industry with news clips, photos, and general news; mining under the Ohio River; and pumps expelling thirty-five hundred gallons of water per minute.

"The Tolu Fluorite Statue," web, Mississippian Indian art, http://lithiccastinglab.com/gallery-pages/2011junetolustatuepage1.htm.

Trace, R. D. "Illinois-Kentucky Fluorspar District," in *Geology and Resources of Fluorine in the United States*, edited by Daniel R. Shawe (Washington, D.C.: U.S. Government Printing Office, 1976), pp. 63–74, https://pubs.usgs.gov/pp/0933/report.pdf.

Weller, J. Marvin, Robert M. Grogan, and Frank E. Tippie. *Geology of the Fluorspar Deposits of Illinois* (Urbana: Illinois State Geological Survey, 1952), p. 14, pumps expelling thirty-five hundred gallons of water per minute; p. 105, thirty-four-foot-wide vein.

"Diamond Drilling, 9 Acres Mine." Powered by a steam engine, the diamond drill consists of a hollow tube with several diamonds on its outer circumference and several more on its inner circumference. The drill cuts through rocks to provide core samples of what lies beneath this Kentucky site. Courtesy Ben E. Clement Mineral Museum, Marion, Kentucky.

At Kentucky's Holly Mine, Ben E. Clement, mine owner and operator, shines a lamp into the Jones Vein, a vein known for its high-quality fluorspar. The writing at lower left reads "Jones Stope"—a stope being a mined-out area. A major figure in fluorspar mining, Clement is also shown on pages 37 and 38. Courtesy Ben E. Clement Mineral Museum, Marion, Kentucky.

A Kentucky miner, wearing a soft cap with an open flame (*see arrow*), works under roof and rib supports of sawed lumber and a split tree. The original caption—"Holly Mine Original Vein near Open Cut 1920"—suggests this part of the operation began as an open-pit, or surface, mine. Courtesy Ben E. Clement Mineral Museum, Marion, Kentucky.

A *log washer* is powered by a steam tractor (*right*) while a shovelman feeds raw ore mixed with dirt into the washer. The unusual name comes from the first such washer, whose inventor attached iron paddles to rotating logs in a water-filled box to clean dirt from ore. The photo was taken at Kentucky's Davenport Mine on February 23, 1931. Courtesy Ben E. Clement Mineral Museum, Marion, Kentucky.

Mules have the right-of-way over autos in this 1924 meeting near Kentucky's Big 4 Mine. Wagons carried the fluorspar to rail centers and to the Ohio or Cumberland Rivers, where it was shipped to end users such as steel mills. Courtesy Ben E. Clement Mineral Museum, Marion, Kentucky.

"Incline. Rosiclare, Ill." Mine cars carrying fluorspar followed these tracks down the incline before emptying into Ohio River barges. Marginalia indicates the site was located at the end of Main Street circa 1905. Courtesy American Fluorite Museum, Rosiclare, Illinois.

An Illinois miner, with a "wet-type" drill (*left*) and several drill bits, poses in front of a mine face into which he will bore holes for the placement of dynamite. His hat and slicker will partially protect him from the water spray used to control dust. Different colors of the face indicate any of several minerals of varying purity may be present—fluorspar, lead, zinc, barite, and others. Courtesy American Fluorite Museum, Rosiclare, Illinois.

Men and machines fill box wagons with processed fluorspar at the loading facility of the Klondike Mine in Kentucky. Mules were used in rural areas in both Illinois and Kentucky until well into the 1950s, when many dirt roads were surfaced with gravel. Courtesy Ben E. Clement Mineral Museum, Marion, Kentucky.

This tally board includes pegs on the numbers 6, 40, and 200, indicating 246 loads of ore passed this way. Courtesy American Fluorite Museum, Rosiclare, Illinois.

A man and a boy stand at an open shaft of a family operation where power comes from a hand-operated windlass. A winter cap on the tree, at left, suggests a third worker may be in the shaft. Information filed with the photo— "Prospects Shafts Miller"—indicates this is a prospecting venture in search of mineral wealth. Courtesy Ben E. Clement Mineral Museum, Marion, Kentucky.

Fluorspar Mining

This no-frills operation shows the headframe—the structure that brings ore to the surface—and an ore car on tracks ending at the cleaning area in the center. The photo was taken in Kentucky at the Susie Beeler Mine circa 1916. America's entry into World War I in 1917 accelerated the search for fluorspar. Courtesy Ben E. Clement Mineral Museum, Marion, Kentucky.

Gravity sends an ore car down a trestle to a cleaning or loading facility at the Marble Mine near Crider in Caldwell County, Kentucky, in 1916. Located on the edge of larger fluorspar deposits, this county was also a major producer. Courtesy Ben E. Clement Mineral Museum, Marion, Kentucky.

Hand-scooping fluorspar into barrels occurs at a Kentucky mine. The presence of only two workers and the use of barrels suggest this is a family mine. Small mines without cleaning facilities often sold their ore to larger mines. Courtesy Ben E. Clement Mineral Museum, Marion, Kentucky.

An important player in the industry, the Kentucky Fluorspar Company was eventually sold to Alcoa, a major manufacturer of aluminum. Courtesy Ben E. Clement Mineral Museum, Marion, Kentucky.

Fluorspar Mining

Miners watch as their large, air-driven, column-mounted drill prepares an entryway. This drill performed well, but its setup was time-consuming. Courtesy American Fluorite Museum, Rosiclare, Illinois.

Mining near the Ohio River carried the threat of flooding in underground mines, and pumps capable of expelling thirty-five hundred gallons per minute were available. At ease in their pumping station four hundred feet underground, miners Charlie Gebauer, Luther Hicks, and John Dalton took time out from their duties to pose for a photo at the Rosiclare Lead and Fluorspar Mining Company circa 1917. Courtesy American Fluorite Museum, Rosiclare, Illinois.

At Kentucky's Hodge Mine in 1916, surface mines were safer and more economically advantageous than shaft mines, but few active operations would have won contests for neatness. Some surface mines ceased operations from December through March because of wet working conditions and impassable roads. Courtesy Ben E. Clement Mineral Museum, Marion, Kentucky.

A variation of a photo in the introduction, the prominently displayed picks and shovels in this image affirm the labor-intensive nature of early mining, performed here with lighting from open-flame lamps with reflectors. Water from the hose was used for dust abatement in the drilling process and to "spray down" piles of dusty ore and headache-inducing gas from blasting. These Kentucky miners at the Big Four Fluorspar Mine are Fred Cooper, Claude Cooper, Orville Croft, Hosie Croft, Dewey Corn, Taylor Lynn, and Chester Robinson. Courtesy Ben E. Clement Mineral Museum, Marion, Kentucky.

One miner performs old-fashioned handpicking to remove unwanted materials as his partner empties the contents of a car into a chute. Courtesy American Fluorite Museum, Rosiclare, Illinois.

Happiness is good friends—and a railroad car full of ore from your own mine. Mine operator and successful entrepreneur Ben E. Clement (*right*) is shown with friends in 1920. Courtesy Ben E. Clement Mineral Museum, Marion, Kentucky.

Fluorspar Mining

In Kentucky, a headframe built of heavy timbers towers above the Baily Shaft of the New Klondike Mine. Later, most headframes were made of nonflammable metal. Courtesy Ben E. Clement Mineral Museum, Marion, Kentucky.

Photos from Illinois and Kentucky, 1905–1995

39

A worker on a Cumberland River barge watches as a truck driver positions his load before dumping at the Dycusburg Tipple circa 1937. The Cumberland offered access to the Ohio River and inexpensive shipping to markets, as seen in the next photo. Courtesy Ben E. Clement Mineral Museum, Marion, Kentucky.

Fluorspar Mining

Driver "Possum" Henry empties cleaned fluorspar into a Cumberland River barge circa 1937. Each barge could carry 550 tons. Courtesy Ben E. Clement Mineral Museum, Marion, Kentucky.

With its wooden headframe and outbuildings perched on a Kentucky hillside, Reed's Mine #1 was typical of many fluorspar operations—but with an important difference: this mine was only a few feet from a township road and telephone service. Courtesy Ben E. Clement Mineral Museum, Marion, Kentucky.

A postcard depicts the "Lafayette Fluor Spar Mill, Marion, Ky." The owner of the Lafayette processing plant, U.S. Steel, was a major producer, refiner, and consumer of Crittenden County's abundant fluorspar deposits. Courtesy Ben E. Clement Mineral Museum, Marion, Kentucky.

American Fluorite Museum, Rosiclare, Illinois.

RESTRICTED
AREA
BY ORDER OF
WAR DEPARTMENT

Fluorspar was vital in the making of steel for armor, aluminum for airplanes, and the processing of uranium for the atomic bomb—hence this World War II sign at a large mine in Rosiclare. Courtesy American Fluorite Museum, Rosiclare, Illinois.

A miner rests his arm on a large metal bucket, or "can," next to a loading machine in a mine owned by the Mahoning Company in the early 1940s. Fluorspar workers were exempt from the military draft in World War II because they were employed in an industry vital to national defense. Photo "contributed by the Mining and Mineral Industry" in *Mines and Minerals in Illinois: An Industrial Romance*, compiled from records and surveys of State Department of Mines and Minerals (Springfield, IL, ca. 1945).

Kenny Clanton uses a jackleg drill to prepare blast holes before inserting dynamite in a face at the Hill-Ledford Mine in 1958. His spacious work area was typical of many of the mines near Cave in Rock in which fluorspar formed in high and wide horizontal beds. Courtesy American Fluorite Museum, Rosiclare, Illinois.

"Headframe & Crushing Plant at Minerva Mine #1." A steel headframe towers above the crushing plant where raw ore will be milled into smaller and easier-to-process sizes, yielding fluorspar, lead, zinc, and other minerals. Courtesy *Hardin County Independent*, Elizabethtown, Illinois.

Surrounded by the steel beams of the headframe, underground workers emerge from the *cage*, an industrial elevator that lifts and lowers miners and materials. Courtesy Ben Gelman Collection, Special Collections Research Center, Morris Library, Southern Illinois University Carbondale.

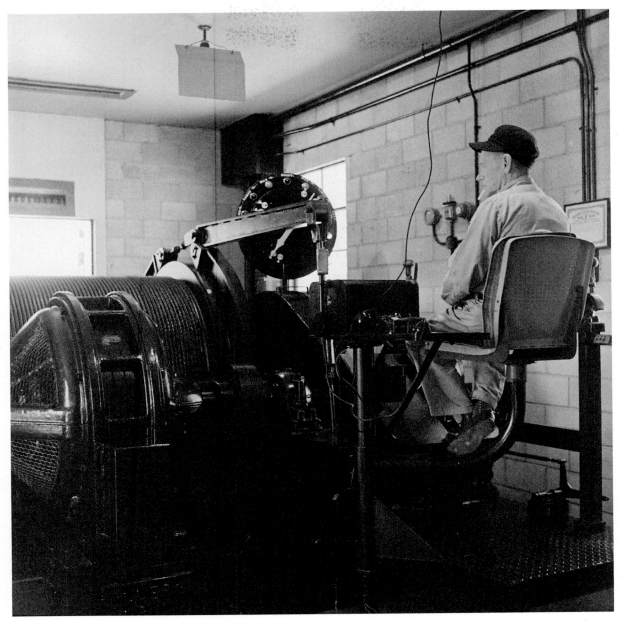

The hoistman at Illinois's Hill-Ledford Mine raises and lowers the cage with the cable and wheel at his left. A standby diesel generator is available in the event of power failure. Courtesy Ben Gelman Collection, Special Collections Research Center, Morris Library, Southern Illinois University Carbondale.

CODE OF HOISTING SIGNALS

1 BELL — STOP IF HOIST IS IN MOTION

2 BELLS — LOWER CAGE

3 BELLS — PREPARE TO MOVE MEN
(FOLLOW WITH 1 BELL FOR
HOIST OR 2 BELLS FOR LOWER)

STATION SIGNALS

1-2 — COLLAR

2-1 — 1ST LEVEL

2-2 — 2ND LEVEL

2-3 — 3RD LEVEL

2-4 — 4TH LEVEL

3-3 — RELEASE CAGE

TO MOVE CAGE FROM ONE LEVEL TO ANOTHER
RING THE STATION SIGNAL FOR THE LEVEL
DESIRED FOLLOWED BY THE CORRECT
HOISTING SIGNAL.

TO CALL CAGE RING STATION SIGNAL.

This Illinois mine used bells to enable the hoistman and underground workers to communicate. For example, two bells followed by four bells would send the cage *down* to the *fourth* level, typically about four hundred feet underground in mines where fluorspar was located in vertical veins. Courtesy American Fluorite Museum, Rosiclare, Illinois.

A worker waits as ore from a chute on an upper level cascades into one of several cars, or *trams* (small wheeled vehicles that run on rails).

A miner dislodges small pieces of fluorspar ore that remain in a chute. The photo also suggests how easily a worker might inhale fluorspar's companion material silica, the source of the serious lung disease silicosis. © 1944 University of Illinois Board of Trustees. All rights reserved. Used with permission of the Illinois State Geological Survey.

William Vaughn maneuvers an underground dragline called a *slusher* among wooden roof supports in Hardin County's Crystal Mine circa 1950. The hand controls move the cable, which pulls a bucket of ore toward him for emptying and then returns it for another load. The cable may turn on a pulley anchored to a steel hitch in the wall or one anchored by other means, as seen in the next photo. Courtesy American Fluorite Museum, Rosiclare, Illinois.

The helmet lights of J. O. Keef and Ben Taylor are visible near a pulley and a cable on which a slusher bucket moves back and forth in the Blue Diggings Mine. Courtesy American Fluorite Museum, Rosiclare, Illinois.

The operator of a slusher on crawlers moves an empty bucket (*see arrow*) before filling it and pulling a load of loosened ore, or *muck*, up the ramp at his left, as seen in the next photo. This photo illustrates the room-and-pillar method of mining in which a room is mined out and pillars are left as supports.

A loaded slusher bucket at the top of the ramp empties into a waiting truck as a helmeted driver observes. This truck is equipped with a scrubber on the running board to clean engine exhaust. © 1954 University of Illinois Board of Trustees. All rights reserved. Used with permission of the Illinois State Geological Survey.

A guide has removed his helmet light for a closer examination of an ore sample in the Ozark-Mahoning Mine as a tour group observes. Courtesy Ben Gelman Collection, Special Collections Research Center, Morris Library, Southern Illinois University Carbondale.

　　　　　　　　　　　　　　　　　　　　Fluorspar Mining

A large piece of dark blue fluorite from Illinois's Denton Mine rests on an antique cart in the American Fluorite Museum in Rosiclare. Another popular source of collectibles was the Ozark-Mahoning Mine. It allowed miners to carry out of the mine (and sell) as many attractive fluorite samples as they could fit in their lunch buckets. Officials later noticed that some miners began carrying their lunch in ice chests and hefty coolers. Courtesy American Fluorite Museum, Rosiclare, Illinois.

"Underground 9 Ton Diesel Scoop Loader (Note: Overhead Roof Bolting)." The driver of a scoop loader makes his way through an immense room-and-pillar mine near Cave in Rock. The side-mounted steering wheel allows the driver to turn his head and see clearly in forward gears or reverse. Roof bolts are visible above the driver (*see arrow*). Courtesy *Hardin County Independent*, Elizabethtown, Illinois.

"Underground Lakeshore Loader (9 Ton per Scoop) Minerva Oil Company's Mine #1." An articulated scoop loader—one that can bend in the center—backs down a corridor in Minerva Mine #1. Capable of carrying nine tons, this model required a special permit to operate because of its diesel engine emissions. Courtesy *Hardin County Independent*, Elizabethtown, Illinois.

"Modern Gardner-Denver Production Face Driller (Note Roof Bolts)." The operator of a Gardner-Denver twin-boom face drill and an assistant pause before drilling holes for blasting at Minerva Mine #1. Roof bolts are visible at the top of the picture. Courtesy *Hardin County Independent*, Elizabethtown, Illinois.

"Production Drill at Minerva Oil Company's Mine #1." The hydraulic hoses allow the operator to maneuver both booms, which drill holes before blasting. In mining's early days, holes were created manually by a worker with a four-pound hammer and a handheld drill. Courtesy *Hardin County Independent*, Elizabethtown, Illinois.

Namesake of the Illinois mine, the Roman goddess Minerva appears ready for battle with corporate America, or with the nation's enemies. Courtesy American Fluorite Museum, Rosiclare, Illinois.

CORE DRILLING JAN/1995
ON DEON ROBINSON FARM,
HARDIN COUNTY, ILLINOIS
L→R ERIC LIVINGSTON, HAROLD MILLER
DON TURNER, NORMAN CUBLEY,
SUSAN PRATHER

Marking the end of an era, a core-drilling crew from the Ozark-Mahoning Company searches for fluorspar in Hardin County in January 1995. The year would mark the end of large-scale fluorspar production in the Illinois-Kentucky Fluorspar District. (*Left to right*) Eric Livingston, Harold Miller, Don Turner, Norman Cubley, and Susan Prather are shown. Courtesy American Fluorite Museum, Rosiclare, Illinois.

In Memory of the Miners Who Have Given Their Lives in the Fluorspar Mines, David Seagraves's 2017 double statue, depicts a wife handing her husband a dinner pail and an extra lamp as he shoulders a drill and goes to work. A baby bottle protrudes from a pocket in her sweater, a reminder that more is riding on his shoulders than just a drill. Located on Rosiclare's Main Street near the Ohio River, the statue memorializes the era when the town was billed as the "Fluorspar Capital of the World." Photo by Bethany Belford.

Herbert K. Russell, formerly the executive director of college relations at John A. Logan College, is a literary scholar and Illinois historian who has been a college teacher, editor, and writer. He is the author of *Edgar Lee Masters: A Biography* and *The State of Southern Illinois: An Illustrated History*, and the editor of *A Southern Illinois Album: Farm Security Administration Photographs, 1936–1943*; *Southern Illinois Coal: A Portfolio*; and *The Enduring River: Edgar Lee Masters' Uncollected Spoon River Poems*.

 *A Shawnee Book*

Also available in this series . . .